I0476451

Skin cell research in Miami

Skin cell research in Miami

THE MICROANATOMY OF HUMAN SKIN

Denis E. Solomon PhD

Copyright © 2016 Denis E. Solomon PhD
All rights reserved.

ISBN-13: 9781515226758
ISBN-10: 1515226751

Foreword

THIS BOOKLET IS A TRUE account of a British scientist who was granted visiting rights to a major American University Medical School during the summer of 2001 to conduct a few simple cell culture experiments involving human skin cells. No one including me, envisaged a major outcome from a simple research project.

I changed long established skin cell culture techniques and the results were hard to interpret. There were no published scientific or medical papers to guide me to an explanation.

The board-certified dermatologists on the University of Miami School of Medicine's staff thought they had learned everything about the human skin at Medical School, on post-graduate medical courses or after, during their clinical practice. However, my cellular photomicrographs, when I correctly interpreted them and offered up a scientific thesis after more than a decade of supreme effort, exposed a cross-fertilisation of scientific/medical disciplines and a mathematical correlation to the human epidermis which was wholly elucidated.

They provided an updated insight into subcutaneous wound healing and the treatment of burns by a new understanding of cellular interactions between the epidermis and dermis as well as cellular configurations of both epidermal and dermal cells.

I have used layman's terms throughout the script only citing scientific terms when they were absolutely necessary. Please read carefully. The numbered superscripts in the manuscript refer to my scientific publications which are fully cited in the Appendix.

My thanks to the *Open Journal of Regenerative Medicine*, the *International Journal of Experimental Pathology (London), the Journal of Developmental Biology and Tissue Engineering* for their kind permission to reproduce copyrighted Figures and a Table from my published papers; these are numbered exactly as they appeared on publication. It should be appreciated that some Figures containing cellular photomicrographs were previously copyrighted in my granted patents prior to the publication of scientific reports.

Denis Solomon PhD (Lancaster).
Wigan, England.
June 2015.

ON THE ROOFTOP OF THE Miami Heart Institute (MHI) on Meridian Avenue, Miami Beach in 1991 was an open air patio equipped with elegant weather-proof patio furniture. The western view was panoramic looking towards the skyscrapers that dotted the Miami skyline. Indoors was a staff restaurant catered by the Marriott Hotel. The patio became my destination when I was wanted to escape my office which was situated in a converted elevator shaft. I found it claustrophobic; my desk, a large comfortable executive leather chair and a filing cabinet filled all the available space. Having coffee on the patio at midmorning and mid-afternoon broke up my working day into two halves. I would take published scientific reprints and browse through them there.

I had arrived after an inexplicable delay. The paperwork authorizing my working visa, which I eventually collected at the USA Embassy in Grosvenor Square, London had apparently been sitting around there on some member of staff's desk. He or she was prompted by the legal eagles of the Miami Heart Institute's Administration staff to send for me to get the ball rolling. I was to be a Research Fellow in MHI's Research Division, studying blood clotting factors, but more importantly to set up human endothelial cell lines from human aortic tissue. It had been arranged for me to collect segments of vascular blood vessels from the Department of Transplantation Surgery on the campus of the University of Miami School of Medicine. I was to wear a beeper which would prompt me to the fact that available cardiac tissue was available for collection. However, on arrival my first job was to set off there to show my face, to introduce myself and to leave capped sterile plastic urine containers containing nutrient cell medium and a mixture of antibiotics plus a small supplement of an antifungal agent. I was soon to learn that I could be called out at any time of the day and night and at weekends, I could not go out to dinner nor have any alcoholic drinks, because I needed to drive a designated MHI staff vehicle on my collection round.

Once human tissue is removed from the human body the pathological decay process starts. It was essential for me, regardless of time of day, to make the collection soonest, get back to the MHI laboratory and start the isolation of the vascular cells. I would sit at a laminar flow hood which is part and parcel of a cell culture laboratory. The flat, working surface of the interior is made of stainless steel. I would switch off the ultraviolet light and put on a normal light. I would place the sterile capped urine container within, thoroughly wash and scrub my hands and put on a white laboratory coat, a pair of plastic safety glasses and finally a pair of surgical disposable gloves. I would unwrap a sterile pink dental plate from its protective paper tissue, place the blood vessel on it and with a pair of scissors cut it lengthwise opening it outwards using 4 pins to secure each of 4 edges to the dental plate. Pinned out, the intimal lining of the blood vessel is exposed. This is the surface directly in contact (or intimate...hence its name) with the blood stream within the living human body. The blood vessel segment resembles an extra large white fleshy straw made of connective tissue. With a surgical scalpel with blade attached, I would gently scrape the blood vessel surface and gently, manually swirl off the cells into nutrient medium contained in a sterile plastic centrifuge tube sitting in a test tube rack.

The cells from the intima (or intimal lining) are called endothelial cells. I preferred to use a 25 ml. plastic centrifuge tube because of its broader neck (compared to a 10 ml. tube). The centrifuge tube was screw capped, centrifuged for about five minutes and subsequently, a visible cell pellet would be seen at the bottom of the centrifuge tube. The pellet would be re-suspended by using a plastic pipette and 3-4 ml. of warm nutrient medium and gently drawing it and the nutrient medium, up and down inside the pipette. This cell suspension was carefully layered on the bottom or working surface of a sterile culture flask (coated with 0.2% gelatin); the flask was partially screw capped and put into an incubator. 5% Oxygen and 95% Carbon dioxide comprise the routine gas mixture permeating within the incubator.

To recap, on entering the laboratory, ceiling lights are switched on; the water bath temperature is checked to determine it is holding at 37^0 Centigrade and a bottle of nutrient medium is taken from the fridge and put into the water bath. The scraped cells cannot be put into cold nutrient medium. The change in temperature can kill them. They must be placed into nutrient medium at 37^0C. Nutrient medium is commonly called cell medium. This is normally red in colour and contains a large cocktail of nutrients, which would not fit on an A4 page, if listed. An incomplete list would consist of inorganic salts, vitamins, amino acids, Earle's salts, L-glutamine, Sodium bicarbonate and other special supplements. Culture media are specially formulated and have specific names. I would normally use Medium 199 to which a supplement, endothelial cell growth factor as well as antibiotics and for initial isolation only, an aliquot of an antifungal agent would be added. Cells have specific nutrient needs. Human endothelial cells will not grow (undergo mitotic division) without a growth supplement being present in the culture medium. In my past, these growth supplements were not pure. In 1991, I used 50µg/ml. of a crude extract of endothelial cell growth supplement, which I made myself in MHI's laboratories. Aliquots were put into vials and deep frozen for later use. Nowadays, some companies specialise in manufacturing growth supplements for different types of cell growth.

Endothelial cells are called anchorage-dependent cells. They anchor themselves onto a surface or substrate. This is in contrast to cells that grow in suspension and do not 'drop anchor'. To fully explain a cell and its secreted ECM, I must draw an analogy of a fried egg *'sunny-side up'*. If the yolk is the cell, then the white is the ECM. Millions of cells will naturally secrete a microscopically-thin sheet of ECM underlying the cells. After the cells are detached, a cell-free, slightly greyish-white flat surface is observed under the microscope. It cannot be properly photographed with an ordinary camera, because it has no distinguishing features at low magnitude.

After a week or more (with fresh nutrient medium being added every 2-3 days) the cells would have multiplied to such an extent they would wholly cover the working surface of the T-25 flask. This is called 'confluence' and the population of the cells would be approximately 6 million. When human umbilical vein endothelial cells (HUVECs) are left in the incubator for a few days after reaching confluence, they can show 'sprouting'; tiny stringy outgrowths which are the precursors of vascular capillaries (See Figure 1 on the next page). HAECs at confluence do not have the tightly knitted cobblestone morphology of HUVECs. Because they have been isolated from a major artery, the cells are larger in size; hence the cobblestone morphology is a bit broader. They do not display 'sprouting' post-confluence. Specific tests would follow to positively identify them as endothelial cells. Specific cell markers for endothelial cells only, would be used. The population of cells in the flask would be split into 2 (using 5mM EDTA/CMF-DPBS at pH 7.4) and re-plated. Those 2 flasks (with a gelatin coating) would be labelled with an ink marker as P_1 (indicating passage 1). Alternatively, the total population of cells could be plated in a T-50 flask (double the size of a T-25 flask) or plated into several plastic petri dishes (again with a gelatin coating).

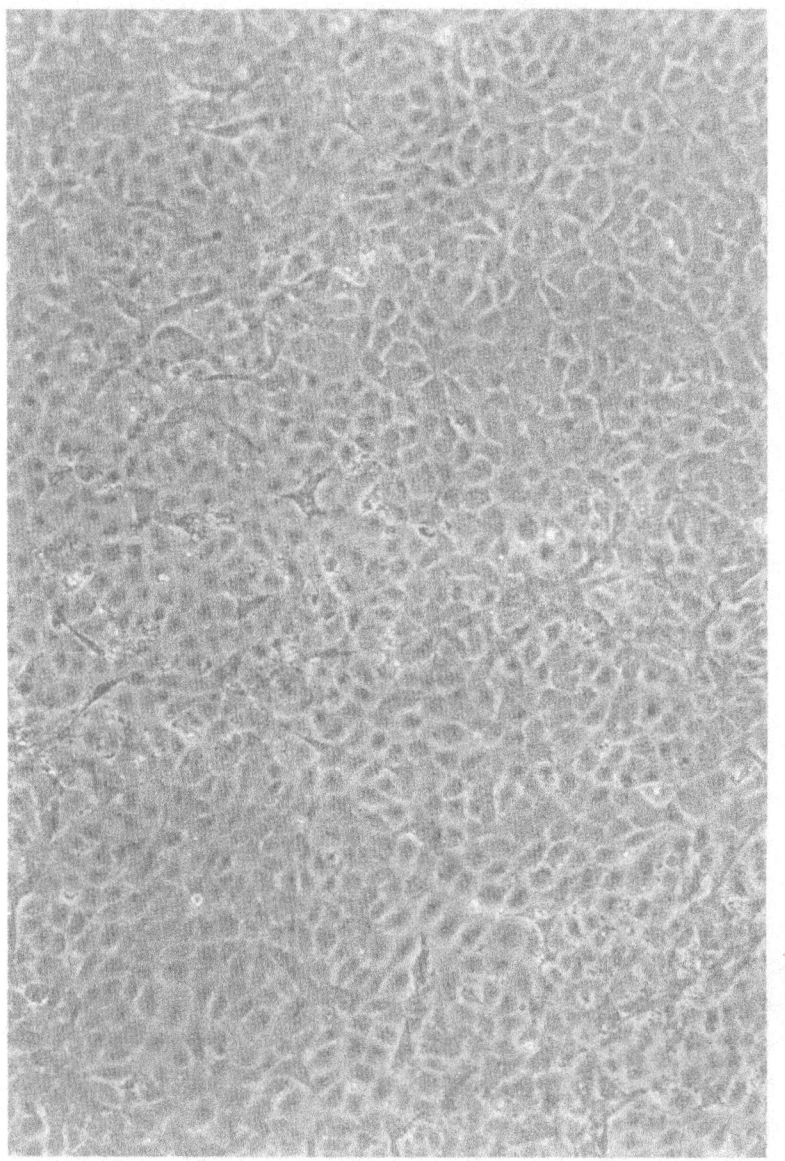

Fig. 1. A post-confluent HUVECs monolayer (6th passage) displaying cobblestone morphology before induced detachment. HUVECs were isolated by collagenase perfusion and grown on 0.2% gelatin/ DPBS substrate. Phase-contrast micrograph. 3mM EDTA /(CMF) DPBS aliquots were added until a cell free ECM was exposed. × 10.

[From reference 1; reproduced with permission]

The pelleting of cells by centrifugation was an established procedure in cell culture. The centrifuge tube containing the cells would be put in a desk top centrifuge and spun for five minutes at a low speed. I always had doubts/ second thoughts about this methodology but was not in a senior position to question it. Many years later, when I had a free hand and was working on my own, on human skin cells, I changed this procedure. It had enormous consequences. But I am getting ahead of myself. Back to MHI.

Every working day was filled with the loud ticking sound of a haemocytometer as it counted the seconds of a test solution mixture to coagulation. It would stop ticking when the blood factors had clotted. The constant ticking sound drove me to distraction. I was happy to escape to the roof top patio. I must say working conditions were pleasant. However, after work, the Miami traffic rush hour would be in full swing. Trying to drive anywhere needed the patience of a Saint. Even back then, Miami drivers had a certain reputation. I knew the major road arteries in Miami Beach but didn't know Miami at all, where spoken Spanish was *de rigueur*. During my initial stint at UM, I lived on the Beach, commuted by public buses and witnessed the start of its renaissance to the now famous SoBe (South Beach). I couldn't get an after dinner drink or relax just in case the beeper went off. The blood vessels were intriguing. One did not know what one would see, on cutting them open. Some would be atherosclerotic, meaning the scraping would yield a cobweb almost devoid of endothelial cells; other intact intimal surfaces had what looked like blisters on the intimal lining. Only the age of the late patient was supplied. The vascular vessels were supple and white when at a young age. The ageing process caused the vessels to go slightly beige in colour (like human teeth) and lose a degree of suppleness. I should say here that most cardiac tissue segments came from the descending portion of the human aorta. On seeing a clear intimal surface of a deceased patient aged say, 55 years of age, one wondered about their diet, their lifestyle et cetera. However, those questions could never be asked or answered.

I had got acquainted with my MHI supervisor (Head of the Research Division) while working in an exclusively endothelial cell laboratory at the University of Miami School of Medicine (UM). There I was put in charge of creating human endothelial cell lines. The laboratory protocol there did not use the methodology of using enzymes on tissues, collecting the cells, centrifuging, re-suspending and then plating the cells in a culture dish. Silently, I personally agreed with this methodology. I learned, without any supervision to scrape human tissue with a surgical scalpel and blade, getting better, meaning developing a lightness of touch when scraping, as time went by. My later boss at MHI was interested in acquiring some human aortic endothelial cells from me. He was supplied after I obtained permission to give a small population to him.

I was interested in what the cells secreted. The secretion of the underlay is called the extracellular matrix (ECM). When the cells reached confluence, one could not use a cell scraper to mechanically remove them because of the 0.2% gelatin flask coating. The cells would mix in with the gelatin and would not attach to a new substrate ; 'not sit down', in cell culture jargon. Enzymatic digestion would have to be used, normally a mixture of Trypsin/EDTA*.

On the roof top patio one day in 1991, I came across a published paper which reported similar cell secretions from two unrelated tissues. I made the mindful, impulsive leap that their ECM must be similar and therefore one should be able to transplant cells from one tissue to the ECM of a different parent tissue. The two tissues were the human aorta and the human umbilical vein (contained within the human umbilical cord). The endothelial cells are named human aortic endothelial cells (HAECs) and human umbilical vein endothelial cells (HUVECs), respectively.

*Footnote: *EDTA= Ethylene diamine tetra-acetic acid; a chelating agent that binds divalent cations like Ca^{2+}*

My supervisor was due to be overseas at a conference and I sought permission to do some simple experiments to prove my thesis. By the time he returned to work, I had done the experiments and written a short paper. I had transplanted HAECs onto the prepared HUVECs ECM and proved that they were fully functional after the transplant. I had also transplanted HUVECs from the vein of an umbilical cord onto the prepared HUVECs ECM from another umbilical cord. The endothelial cells rapidly attached, proving a similarity in ECM composition in different cords.

It was a stipulation that any proposed publication should be submitted to an MHI publication committee of senior medical doctors. My supervisor had reservations about the value of the work. I decided to make personal representation and ran into the late Dr. Benson near his office. My words were that I was willing to defend my paper. He surprised me by replying that I should go ahead and publish. My name was given as the sole author. My supervisor was not keen on coming aboard as co-author. I decided on a top journal, thinking that if I were to be shot down in flames, I should do it with some style. To my surprise, the paper was readily accepted for publication with very minor corrections (See Figure 3 on the next page, reproduced from reference **1**); which shows the intimal lining of an aortic tissue segment, (containing human aortic endothelial cells, HAECs) obtained by gently scraping with a surgical scalpel and blade, which has attached to the HUVECs ECM, and the outgrowth of HAECs onto the HUVECs ECM.

This was some justification for changing my research focus to cell culture and the study of cells. I had been introduced to cell culture of human umbilical vein endothelial cells (HUVECs) by Professor Gerald Berenson MD at Louisiana State University (LSU) Medical Centre in New Orleans in the mid-1980s. After my tenure at Oxford was over, I had searched the published literature for someone who was studying something I would find interesting. I arrived out of the blue and asked to see him. He said my employment was possible if I could wait three months until a grant was funded. It wasn't a position I would have normally taken, but I was there and the sensible thing was to wait.

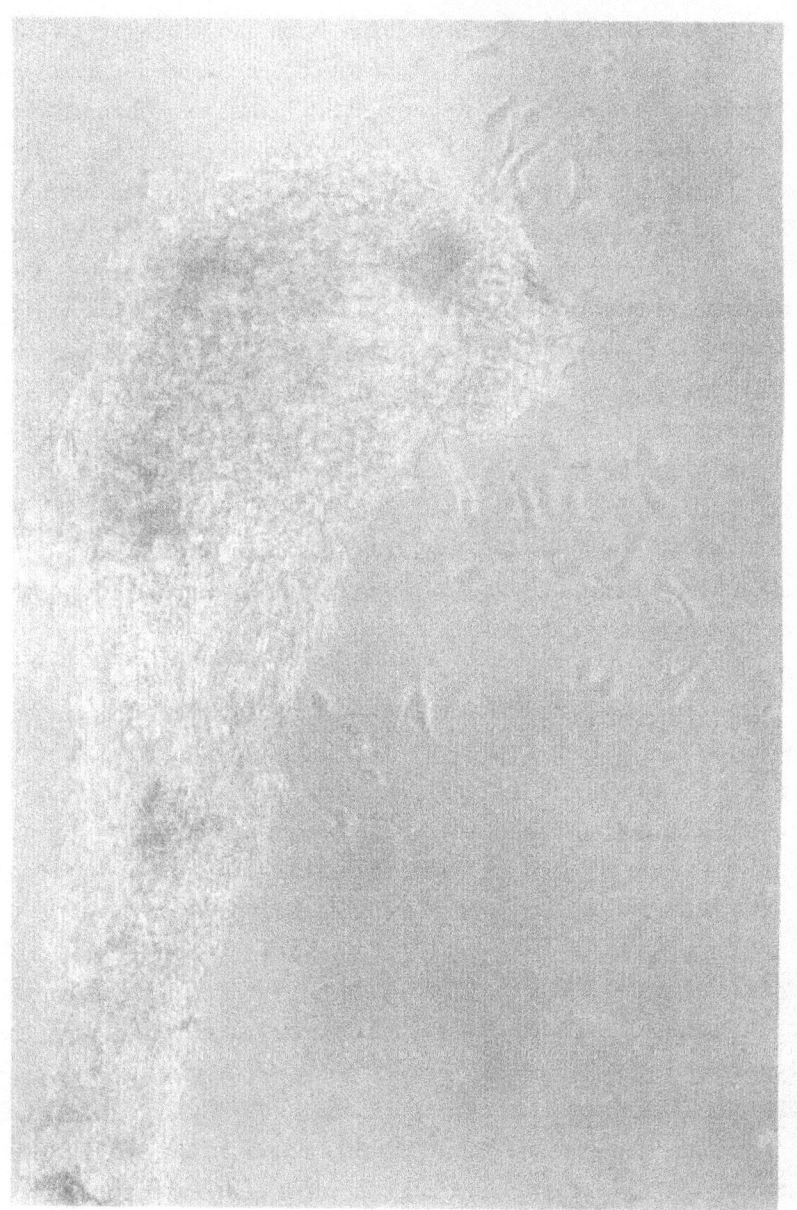

Fig. 3. Outgrowth of HAECs from an ECM-attached endothelial sheet. This occurred within 16 hours of scraping the aortic segment. Phase-contrast micrograph. × 10.

[From reference 1; reproduced with permission]

My introduction to cell culture was dangerous. The technique in use was as described in a published 1973 paper by Jaffe EA, Nachman RL, Becker CG and Minick C. in J. Clin. Invest. **52**, 2745-2756 entitled, '*Culture of human endothelial cells derived from umbilical veins*'. The umbilical cords were obtained from Charity Hospital, a public hospital which sits between the two medical centres of Tulane University and Louisiana State University. I knew little about AIDS HIV. Neither did my co-workers. There was a group of us and Dr. Berenson when he could spare the time. We couldn't attach the metal cannulae called male and female to either end of the cord (to enable saline perfusion followed by collagenase perfusion, when the saline effluent out of the bottom female cannula had become clear; not coloured by blood) and blood splattered everywhere within the laminar flow hood. My protection was a pair of flimsy disposable gloves! We tried knotting surgical silk sutures to position the cannulae to the cord to no avail. Getting a cord without clamp marks post-delivery was always a problem. At MHI, it took me a month before the nurses at Mount Sinai Hospital on Miami Beach could produce useable material. The cord must be intact for the vein to be perfused. If pinched by clamp marks post-delivery, there could be no assurance that the intimal lining would be wholly intact; in fact the reverse was true. Looking down on one end of an umbilical cord, one sees two short white stalks which are arterial vessels and miniature pinkish 'human lips' comprise the entrance to the vein. The cord, usually greyish in colour, is quite slippery (akin to an eel) and hard to handle. It slides out of gloved hands quite easily. I actually dissected the vein longitudinally along the side of a cord to discover it had a staircase appearance, going sideways down and across.

Years later, while I was working at MHI, I was idly tying the end of a plastic package of sliced bread on the kitchen working surface in my apartment kitchen and happened to glance down at the domestic twist tie tightening the curled, closed end of the plastic bag. They worked a dream in attaching metal cannulae to cords and they could be easily sterilised by autoclaving! A haemostat (a stainless steel medical clamp) maintained the position of the outer tightened twist tie and the inner, inserted cannula at each end of the cord.

After my 1992 ECM paper was published, I searched far and wide for an ECM-designated laboratory but could find none. If a research group was studying endothelial cells, attention was being paid to one component of their multifaceted cell secretions and not the underlay, the ECM as a whole entity. My instincts told me that the ECM was important but how and what was its use? I decided to bet the kitchen sink that some other tissue had similar cell secretions to both HAECs and HUVECs and offer a degree of commonality in its secreted ECM. Which tissue? The library search began in the mid-1990s using the John Radcliffe Medical Library at Oxford and the Calder Medical Library on UM's campus in Miami. I had made a pit stop at Oxford University (at the Nuffield Orthopaedic Centre) as a Research Fellow in the early 1980s, thus I was granted visiting rights to the medical library. I also had a couple of cell consultancies in two different medical departments at UM, each on a 3-month basis, since I had last worked at UM for a year as an International Scholar, thus I qualified for visiting rights there too. The consultancies were cheerful assignments and got me out of UK winter weather.

On the 28th March 1999, I found the link and it was to the human skin. I was in Miami at the time. It was in a published paper by Kramer RH, Fuh GM and Karasek MA (1985) entitled: *'Synthesis of extracellular matrix glycoproteins by cultured microvascular endothelial cells isolated from the dermis of neonatal and adult skin'*. J. Cell Physiol. **123**(1): 1-9. A Chinese proverb states that *'A journey of a thousand miles begins with a single step'* (Lao-tzu, Chinese philosopher; 640 BC-531 BC).

Finding that correlation was the start of an eventful scientific journey. My thoughts gravitated towards wound healing, but I did not know or could not imagine where it would lead me as I followed in dogged thoughtful pursuit. There was an established technique for the cell culture of human keratinocytes, the major cell player of the human epidermis. Irradiated mouse (murine) fibroblasts were being used as an underlay to allow the fibroblasts to secrete their growth factors leading to the growth or population doubling of the keratinocytes. Meanwhile, burns patients had to wait 3-4 weeks for an expensive flimsy keratinocyte graft to be manufactured using his/her own keratinocytes. This is called an autograft; auto- means self. I read the widely used Harvard technique (Rheinwald and Green 1975) for growing keratinocytes with the irradiated fibroblasts which had dominated this research field for decades and declared it a red herring within 10 minutes. Better brains than mine had been working on wound healing since the dawn of civilization and I decided they all had to be making a mistake. But what was the mistake? (See page 38 for the answer).

The human skin was first studied in the year, 1627 by Adriaan van de Spiegel (See Appendix). Remember the Bill of Rights was not signed in England until 1689. It is the largest organ in the human body and consists mainly of the epidermis and dermis. The epidermis does not contain blood vessels and is referred to as an avascular tissue. It is impermeable, that is, it is waterproof. It is extremely thin, thinner than a single ply of toilet tissue. The dermis contains blood vessels. When you cut your finger, and blood seeps, it is from the blood vessels in the dermis, underneath the epidermis. You have cut into the dermis.

When you get a suntan, cells called melanocytes only found within the basal or bottom layer of the epidermis transfers a brown pigment called melanin to the keratinocytes causing the latter to darken and the skin to become tanned in appearance. A basal cell carcinoma, a form of skin cancer, can be caused by too much ultraviolet light from sunbeds or exposure to the sun causing unknown molecular perturbations within the melanocytes.

Wound healing had been at the back of my mind since the late 1980s. Taking a job with a UK Government Research institute, I used to get a lift back to the neighbouring village where I had temporary digs. One afternoon, the friend who offered me a lift had to work late, so rather than walk to the train station and await a train, I ended up in the onsite library. These premises harked back to another era with its polished wood panelling and its tall, wooden book cases. I picked up a book and started to read. It's a pity I cannot remember the title of the book or the chapter I was reading. It stayed at the back of my mind like papers I would later find in my constant online literature searches. Some papers stood out but I could not fit them in my mind and my thoughts at that time. However, they registered on my mind.

Having found the correlation with human skin, I was able to compile a table of cell secretions for HUVECs, HAECs, the dermis, human keratinocytes and human dermal fibroblasts and recognize that there was some commonality in those extracellular cell secretions (See my Table on the next page).

Table 1. Commonality of cell secretions

Human endothelial cells	Extracellular Secretions	References
Aortic, adult vena cava	Type IV procollagen, thrombospondin and fibronectin.	Sage & Bornstein (1982); Fry *et al.* (1984)
Umbilical vein	Type IV procollagen, thrombospondin and fibronectin.	Sage & Bornstein (1982); Fry *et al.* (1984)
Neonatal/Adult dermis	Type IV procollagen, thrombospondin, fibronectin and laminin.	Kramer *et al.* (1985)
Neonatal/adult skin fibroblasts	Laminin, type IV collagen, perlecan, nidogen/entactin and keratinocyte growth factor (KGF).	Woodley *et al.* (1988); Fleischmajer *et al.* (1988)
Epidermal keratinocytes	Type IV procollagen, type IV & VII collagen, laminin, fibronectin and vascular endothelial growth factor/ vascular permeability factor (VEGF/VPF).	O'Keefe *et al.* (1984); Petersen *et al.* (1988); Stenn & Malhotra (1992); Ballaun *et al.* (1995)

© 2002 Blackwell Science Ltd, *International Journal of Experimental Pathology*, **83**, 209–216

210

[From reference 2; reproduced with permission]

At that time, the significance of the upper layer of papillary fibroblasts and the deeper reticular fibroblasts within the dermis had not been studied. The generic term 'fibroblasts' was still being applied to all dermal fibroblasts.

Taken together, this led me to examine every article I could find to read on subcutaneous wound healing, especially the cellular components of the epidermis and dermis. Standard medical practice was studied, their use of antibiotics, immunological considerations (when a patient was so badly burned that keratinocytes from another person was medically used to construct a graft, called an 'allograft'), clinical outcomes of treatments for wounds and burns, haemostasis (the maintenance of physiological functions in a badly burned patient), lung inhalation of smoke (often deadly), the manufacture of autografts by commercial companies, the first generation of cultured skin substitutes which were being commercially manufactured in the 1990s, the use of honey and finally the historical perspective of wound healing from the Greeks and Romans through to the American Lewis and Clark expedition to the evolution of modern day medical treatments. These are just a few items from my long and extensive laundry list.

In essence, I was fully studying a post-graduate medical course in wound healing and the history of science of wound healing in a small way, on my own. I soon realised that no medical school curriculum included cell culture. The follow-on is that no Board-certified dermatologist could take a skin biopsy into a cell culture laboratory and separate the epidermis from the dermis and isolate the epidermal and dermal cells. Needless to say, my self-imposed task took long days and nights of reading, accumulation, percolation through my mind and deep reflection of what I had read. There were times I read photocopies of published papers throughout the night, listened to the acoustical richness of birdsong before dawn and watched the night time shadows being chased away by the brightening in the dawn sky before sunrise. The cost of those photocopies was piling up. I was desperate for funds. I cancelled a UK life insurance policy at a financial loss just to have a small amount of money at my disposal. I asked myself whether that was not a combination of bravado and foolhardiness. However, I was too far down the road. There was no alternative. I had to commit all personal reserves in the hope that things would somehow work out. Two open questions remained a constant niggle. Would I get to the right answer and draw the right conclusions and would these be acceptable to my peers? In other words, would my papers be eligible for publication, after peer review?

I decided to write a UK patent application and finger typed it using the computers set aside for public use in the Miami main branch library located at Government Centre (essentially downtown Miami). I also discovered in the downtown area, the GESU Roman Catholic Church, a wooden structure constructed in 1896 and initially staffed by Jesuit priests from New Orleans. I loved the altar and its stained glass windows along the two sides of the interior. As a former Roman Catholic altar boy, I had the required sense of appreciation. Being an altar boy before I was a teenager and after I became a teenager had instilled in me patience, self discipline and the importance of punctuality. That church would become my refuge in times of personal turbulence, when nothing seemed to be going my way. I would drop to my knees in one of the pews and pray. My visits were always during daytime hours.

At this time, I had digs in Miami and walked the neighbourhoods. In the heat of a Miami summer, sometimes 92^0C with 81% humidity, I decided I had to get fitter to cope with the heat and humidity. I would walk along the sidewalk from the Roads area (SW 17th Road) to Miracle Mile in Coral Gables on Coral Way, one of Miami's main arteries. I would wear a polyester track suit bottom, a colourful T-shirt, a pair of sunglasses, a baseball hat and a pair of Asics running shoes which turned out to be very sensible and comfortable walking shoes. I was never accosted, mugged or felt fear during these long walks. Sometimes, I would also walk back instead of taking the Coral Way bus. My only stop would be at a supermarket to use their water fountain or restroom and never longer than five minutes. The realization that I was in for the long haul with my private research made my personal decision to get fitter, a must-do.

Sometime after, I discovered the human skin link, I learned that Dr. Jeffrey K. Raines once Head of the Miami Heart Institute's Research Division was now a member of UM staff. I spoke to him over the telephone and he kindly arranged to meet with me in his UM office. When we met I informed him that I would like to do some simple cell culture experiments using human skin cells; I did not have an American green card; I was staying in Miami and I had worked in various University departments on various elements of cell culture as a consultant. I would need access to human skin tissue preferably from the operating theatres. He in turn told me that a new Chairman of the Department of Surgery was going to be appointed sometime in the next year or two and nothing could be done until that appointment was expedited. Now, I faced a dilemma. Miami was not a cheap city to live in and cooking for myself saved me money, but the purchase of groceries, toiletries et cetera needed hard cash. I could not afford my rent so I decided to return to England and wait there. In the year 2000, my sister a New Orleans resident for decades, invited me to visit her. Luckily, there was an old friend from my College days in New Orleans as a teenager who invited me to stay. The matriarch of the family was ill in a nursing home and I agreed to stay in her apartment with her young grandson. The latter was into computers in a big way and really opened my eyes to their multiplicity of uses.

I was in contact with Dr. Raines in Miami regularly by telephone. One day, I received a 'phone call inviting me back to Miami. I could gain short-term employment as a Research Coordinator on a clinical study involving the use of some newly manufactured stents. Metallic stents are inserted into ageing blood vessels using angioplasty and a balloon catheter (to position them) to alter the shrinking diameter of the ageing blood vessel permitting an enhanced blood flow, after the cardiac procedure. I readily agreed. By becoming a journeyman scientist in America after my Oxford tenure gave me a much broader scientific pool of knowledge than if I had stayed at home on a single track of scientific research.

By then, real estate developers were in full cry in Miami. Old neighbour-hoods were being torn asunder to put up high rise buildings. My old digs were torn down, so returning there was not possible. I had to find new digs. In the meantime, apartment rents were increasing wildly. Dressing for work required 100% cotton white dress shirts, silk ties, khaki chinos, and slip on shoes. I had to look the part. I often had to insist that my neighbourhood dry cleaners use light starch on my white dress shirts. Trying to iron out the wrinkles on a 100% cotton shirt after domestic washing was near nigh impossible. They had to be dry-cleaned.

This research study went well. Three clinical professors were in attendance, a senior scientist and me, when the surgical procedures were conducted. When my tenure was over, Dr. Raines, the late Dr. SL Hsia (Dermatology and Cutaneous Surgery), Dr. Theodore Malinin (Orthopaedics) got together and arranged for me to use Dr. Malinin's private cell culture laboratory in the Orthopaedic Department on the 7th floor of the Rosensteil Scientific Medical Building (commonly referred to, by UM staff by its acronym, RSMB). I would be an unpaid visiting scientist. It should be noted I had no previous training with the cell culture of human skin cells or with skin studies. Also in the mix was the Head of UM's Department of Biotechnology Transfer, Dr. Gary Margules DSc. These gentlemen were aware of my UK patent application and the original patent letters were given to them for their perusal. I must pay a fulsome tribute to these loyal servants of the University whose main aim was solely the advancement of science. When Dr. Hsia had seen my write-up on the proposed cell culture experiments, including my self-compiled Table, he had crisply declared that they were 'doable and workable'. Scientific peer judgments cannot be any clearer than his statement. Later I learned that these gentlemen were somewhat bemused by my approach. Apparently, it was not every day that a scientist walked off the street and said that he might have some new ideas on wound healing, a complex medical problem for generations. To get these professors to a round table meeting took some doing. Anything from 2-4 weeks was the norm. They were often out-of-state giving lectures at other American Universities or just too busy with clinical work. Dr. Hsia was a scientist, a Research Professor who had been with the University since 1963. His calm quiet counsel was something I always appreciated. He granted me $1500 of laboratory supplies which was mainly used to purchase cell medium and a single endothelial-specific cell marker, used in endothelial cell identity tests.

I started cellular work a month later after the July 4th national holiday. To get to Dr. Malinin's laboratory, I needed card key access down a restricted corridor and a key to open the door lock on his laboratory. I was granted 24 hour access which meant a key card to open the outer back door of the

RSMB (directly opposite the Calder Medical Library Building) during after-hour visits. Dr. Malinin's cell culture laboratory was not modern. It had a single laminar flow hood, a fridge freezer at the side of the doorway and on three sides there were working surfaces with drawers underneath containing laboratory supplies like plastic pipettes, cell culture flasks, petri dishes. Behind the flow hood and near to the entrance door was a wooden table on which lay an old phase-contrast microscope of vintage circa 1970, I thought. There was no camera attached. As I mentioned earlier, I had to wait a month before I received umbilical cords without clamp marks, when at MHI. I got acquainted with the maternity nurses at the Jackson Memorial Hospital (a short walk from RSMB) and remain most grateful for their cooperation. They had to remember to save the umbilical cords and put them in to the capped plastic urine containers I left with them and then telephone me to pick up. Working within the only major Miami hospital with a Trauma Centre meant they were always very busy. They didn't need me coming along and asking for their cooperation with a research project. Walking along those corridors provided a life lesson. Occasionally, I would come across men folk in tears who explained to me when I tried to comfort them with some soothing words that their infant had either been still-born or had died on delivery. I did not encounter happy new fathers.

I started to isolate HUVECs in the laboratory and after confluence, their ECM. To grow these endothelial cells, I decided to use an epidermal cell growth supplement used in cell culture of keratinocytes in Dr Hsia's laboratory. Unchartered territory again, but I observed that the confluence of endothelial cells were obtained in a routine, timely manner as with the use of established technique growth factors recommended for cell culture of human endothelial cells. In my published MHI paper, I had used 3mM EDTA/CMF-Dulbecco's phosphate buffered saline (DPBS) with a pH adjusted to 7.4 on the confluent or post-confluent layer of cells to induce detachment. I adjusted to 5mM EDTA having learned during my cell consultancies (post-1992) that the 5mM EDTA/CMF-DPBS offered a more rapid retraction of the confluent cell layers. To be precise, any culture medium is pipetted from the culture flask and the cell layer is carefully rinsed with an aliquot of

DPBS without Ca^{2+} and Mg^{2+} ions, commonly called Calcium-, Magnesium-free DPBS (CMF-DPBS). This is pipetted off. An aliquot of 5 ml. of 5mM EDTA/CMF-DPBS is added. After about 10-15 minutes in the incubator, clusters of cells which had lifted off the ECM would be seen under the microscope. This procedure left the underlying ECM devoid of overlying cells and wholly intact. After rinsing with DPBS, the flask containing the ECM could be stored in the incubator for a few days.

I was given a summer student as a gopher. He was used to go downstairs to Dr Hsia's laboratory to acquire 5mM EDTA/CMF-DPBS and adjust the pH to 7.4, and to go over to the library to acquire photocopies of published papers of interest to me. He was involved in a major disaster in Dr. Hsia's laboratory when adding the mixture needed to prepare cells for freezing. He added the two constituents consecutively to the cells instead of pre-creating the required mixture of components and killed millions of picture-perfect (as seen under the microscope) HUVECs. The research associate working in Dr. Hsia's laboratory, present at the time, was not acquainted with the laboratory technique used in the freezing of cells. This fact was unknown to me. My specific instructions were disregarded. The cells were only sent down there (to the ground floor) because I had no freezing vials at hand in Dr. Malinin's laboratory. They were being frozen as a ready source of acquiring HUVECs ECMs at a later date.

A senior pathologist on Dr. Malinin's staff had an office a few doors away from Dr. Malinin's cell culture laboratory where I was working and took an interest in my work. Dr. WE Buck conducted autopsies and usually cut off skin tissue segments for his own examination in constructing a medical pathology report. He would give me cadaver skin tissue. Later on, on just a few occasions, Dr. Eton (a vascular surgeon) would call me from his operating theatre and offer fresh skin tissue segments. Again, for their collection and overnight (if needed) fridge storage, I gave out sterile plastic urine containers with cell medium and their supplements of antibiotics and anti-fungal agent. I would cut the skin tissue into 2 inch x 1 inch pieces and rinse these skin tissue segments with DPBS and invert them (meaning epidermis touching the enzyme solution) individually in a small pool of Dispase, an enzyme known to separate the epidermis from the dermis. Literature sources had described conducting this enzymatic digestion at various temperatures using different time periods. There was no standard procedure. I decided to incubate the culture dishes overnight in the incubator at 37°C. The next morning, with a pair of forceps I would carefully peel off the epidermis (essentially separating the epidermis from the dermis) which would curl on itself like a corkscrew because of its thin fragile nature. Seeing that, it was hard to believe that this thin layer of epidermis provided the human body with an exterior waterproof layer.

The peeled off epidermis was found to disassemble when placed in warm DPBS. I christened the disassembled epidermis cellular islands as 'epidermal brown rosettes'. These are intra-epidermal micro patterns as seen under the microscope and appear brown in colour under the phase-contrast microscope. There were no published scientific papers reporting on these islands. I was in new scientific territory. I collected a few islands with a plastic pipette and decided to transplant them onto stored HUVECs ECM. I noticed over a period of time that only a few of them would attach to the ECM. This attachment proved a commonality with the respective umbilical vein endothelial cells ECM and the basal lamina (or basement membrane) underlying the epidermis layer within the human skin. The basal lamina separates the epidermis from the dermal-epidermal junction (DEJ) lying above the dermis, proper. I decided to photograph this cellular structure (the DEJ) now exposed after peeling off the epidermis with a pair of forceps.

What was going on here? Why would a few attach and not others? I decided to photograph them again after noticing under the phase contrast microscope that the attached brown epidermal rosettes shed their cell loads onto the ECM creating a new micro pattern. Photographing them meant placing the culture flasks on a durable plastic tray, taking the elevator two flights down, and walking carefully through the width of the linked Fox Research Building and up the stairway to get to Professor Tseng's laboratory in the Bacom Palmer Eye Institute Building. He had given me his personal permission to use his photographic facilities. His staff operated computer controlled cameras and all photographs of all of my experiments, essentially cellular micrographs, were taken by his polite, cooperative staff. They stored the photographs on floppy disks using a University computer program file which created problems for me in later years, because I could not open the file to access the photographs to study them. I had to go strictly by personal memory. What I did have was single colour photograph printed out on photographic paper of a basal rosette shedding its cell load on a prepared ECM and creating a new micro pattern. I was now working six days a week and it became a test of physical fitness.

I also observed that the brown rosettes gave rise to two different shedding of cell loads when they attached to prepared ECMs. There were two new and different micro patterns (See Figures 1 and 2 in Appendix). There was a single cellular photomicrograph which I interpreted many years later as an inverted attached basal brown rosette with an infantile basal lesion within the shedding load of epidermal cells (See Figure 3 in Appendix). This was never seen before in medical/scientific literature and showed that cell culture could be used for skin cancer diagnostic tests. As I write, I must report that this technique has not been used, probably because there is no way of knowing anyone is harbouring a basal lesion until it manifests itself on the skin surface as a mature lesion.

If I scraped the surface of the now exposed dermis to grow fibroblasts, I noticed under the microscope that the epidermal rosettes would lift off from their co-culture with (HUVECs) endothelial cells and dermal fibroblasts. I guessed that they, as the major epidermal cellular constituent, were out of their normal accustomed milieu and since the rosettes contained a full complement of epidermal cells, it was not the epidermal keratinocytes acting alone. This observation led to my conclusion that the epidermal rosettes did not mix well with cells from the dermis, a vascular tissue (that is one containing blood vessels). As I stated above, the epidermis contains no vascular blood vessels and is termed 'avascular'.

The summer in Miami was hot and humid. I would take a coffee break and walk down to the food court opposite the Bascom Palmer Building to a Cuban restaurant where I would eat a pastelito (filled with guava and cheese) and drink sugarless café con leche. The Cuban waitresses thought that was a travesty of their native custom, because 99.9% of their customers drank the traditional Cuban coffee with sugar.

There were no serious hurricane warnings that summer. I rolled into September now feeling the strain of a six day week. 9/11 arrived with morning television pictures depicting the shock of a lifetime. New York, New York was the epicentre. That night I stood on my apartment balcony and gazed down on a major Miami road devoid of pedestrians and vehicle traffic. Everyone, including me, was at home in a state of mourning. Life however has to go on. The human spirit must prevail and it did.

To isolate dermal fibroblasts, I did not follow established published techniques which consisted of applying an enzyme to dermis which had been minced. In essence, that yield of fibroblasts would consist of a mixture of both papillary and reticular fibroblasts. Dermal fibroblasts have no specific cell marker which can be used for cell identity tests. Hence, there was no known way of differentiating between the two types of fibroblasts in the human dermis.

My personal belief was that I had to isolate these dermal cells using the purest, uncorrupted means. To my mind, that meant scraping the dermis with a surgical scalpel and blade (obtaining a mixture of papillary and reticular fibroblasts or them separately by light scraping of the upper dermis, then following up by deeper scraping of the dermis and seeding directly into two separate flasks), gently swirling off the cells from the blade into an aliquot of cell medium contained in a sterile plastic centrifuge tube. I would seed culture flasks directly. I would not centrifuge disallowing any applied shear forces on the cells. Literature sources had long established that fibroblasts from most tissues, attach easily to the plastic surface of a cell culture flask or petri dish and that had been verified by me during the tenure of my cell consultancies at UM. They readily attached, grew normally in a routine manner and the cell layer at confluence could be retracted as cell clusters of fibroblasts after the application of 5mM EDTA/CMF-DPBS at pH 7.4 and placing the culture flask within the incubator for 10-15 minutes. A dermal fibroblast ECM would be left behind, which proved just as efficient as the HUVECs ECM for basal epidermal brown rosette attachment. Seeing the morphology of my dermal fibroblasts[5], the senior pathologist, Dr. WE Buck asked me to isolate some human periosteal fibroblasts from periosteum, he would supply. This I readily did by scraping and after the cells were admired by both Drs. Buck and Malinin, they were frozen for Orthopaedic Department research purposes.

Taken together, I was tweaking all established techniques and getting some interesting results. There were no published literature references to guide me with my results. I had no understanding of the overall scientific picture, real or imagined. I was not talking about anything, since I lacked knowledge and this led to later remarks that I was somehow being 'secretive', which I considered laughable. Other remarks reaching my ears, suggested that UM medical staff did not understand the nature of my research.

Questioning trauma surgeons about their standard wound healing practices proved to be a diplomatic process. My lasting impression is that if you have a medical degree, there is an air of mutual respect. If like me, you have a PhD degree, there is something in the air which suggests that you might not fully understand the depth of your questions, because you haven't treated patients or don't possess a broader clinical understanding of the issues. I felt that some of them considered my research efforts to be 'wound healing within a test tube'. The clinical professors who I associated with and gave the go-ahead for my project did not have that attitude. The century or more of research experience that Professors Raines and Malinin (plus Prof. Hsia) had acquired with hands-on efforts, gave them a broader, mature outlook on research, be it clinical or non-clinical. Dr. Buck (Orthopaedics), Dr. Eton (Vascular Surgery) and later Professor Tseng (whose polite staff did my photography at the Bacom Palmer Eye Institute) possessed the same qualities. Two years after I left UM, Dr. Hsia confided in me that UM medical staff now recognized the importance of the ECM in the creation of the so-called biofilm in wound sites.

I continued to work, tweaking things as I went along. In peeling off the epidermis after a Dispase digestion, I had accidentally left behind, un-noticed, a very small piece of intact epidermis (because I had no dissecting magnified glass). I was interested in acquiring dermal fibroblasts and had scraped the midsection of that skin segment. I decided to swab the area around the final tissue anchoring point with the scraped skin segment. Putting down a piece of tissue in a tissue culture dish and adding an aliquot of cell medium is tantamount to setting up a classic 'explant' cell culture procedure, in which the cells will migrate away from the anchored tissue, over time. My observations were that from the swabbed area, dermal fibroblasts outgrew first. What was really interesting was that days later, the speck of epidermis migrated away from the skin tissue segment, and the fibroblasts retreated. I presumed they left their ECM behind for the intact bit of epidermis to migrate on, akin to a railway track. On the leading edge of the epidermis, a front line of keratinocytes could be morphologically recognised leading the way along the direction of its migration. Finally, the speck of epidermis transformed itself into a multi-layered island. I did not change the cell medium, realising that something unusual was occurring and not wanting to disrupt the unknown process. Although working six days a week, I realised that I was absent when certain critical observations needed to be made. The process described above was seen by me in bits and pieces until the Labor Day weekend, in September, when I realised it was a continuum and I should not change the culture medium, now a bit cloudy, even though the resulting photographs would not have the required clarity.

I had under my belt, a previous experience of not changing the culture medium and making an unusual observation. In cell culture jargon, after a couple of days, the culture medium in contact with cells is referred to as 'conditioned medium'. Thirteen years earlier, I was culturing HAECs. In the culture dish, on looking down through the microscope, there was an attached cluster of cells in a North East position and an attached single cell to the South West. I am using compass directions to simplify things. To my amazement, the single cell extended a tiny 'antenna' which grew steadily until it touched the cluster of HAECs. I christened that 'antenna', a 'telegraph wire'. I reached the conclusion that the conditioned medium in which the cells lay contained a 'transmitter element' which enabled the specific growth of the 'antenna'. This remains an unpublished observation. Subsequently, I held to a rigid opinion that cell culture could not be properly conducted during the normal working hours of 9am-5pm or 8.30am-6pm. After-hours and weekend work were absolutely essential.

Years later, I realised that the multi-layered island was an *in vitro* structural reversion of the speck of epidermis to the normal epidermal 'stack' of rosettes found in intact epidermis, which I had seen disassembling in warm DPBS, after overnight Dispase digestion. At the time, as with other observations, I did not realise the importance of this experimental procedure. I thought the multilayered island observed at the end of the experiment was simply a matter of 'autoengineering' by un-manipulated epidermal and dermal cells. It was, but much more than that. It could be a paradigm, representative of wounded skin and the involvement of dermal fibroblasts in regenerating the epidermis. In scraping, I had reached down to the deeper reticular fibroblast layer of the dermis. As other scientists published their work on dermal fibroblasts, differentiating the dermis into papillary and reticular fibroblasts, I developed a greater understanding of my own past experiments.

From time to time, there would be academic meetings called in the conference room of Dr. Raines's Department of Surgery. There was general chit chat but I never did understand the real purpose of those meetings. On one of these occasions, Dr. Malinin voiced his opinion that I should go ahead and publish my work[2]. On another day I was called to Dr. Malinin's office in the Orthopaedic Department and formally told that the University would decline any participation in further work. I was never asked what I thought of my experiments and what I would do, going forward, if given the opportunity. I thanked Dr. Malinin for the use of his laboratory. As later events would show that was the day the University threw out the baby with the bathwater; that signalled the end of my tenure as a visiting scientist at UM laboratories.

Before I left Miami, I wrote a paper for Journal submission. I omitted a description of the experiment on 'autoengineering', because it did not fit in with the overall scheme of the paper. Although the University had no rights to my work, it was published under UM's banner. No one had recognized the importance of my work and no one asked to be a co-author. I later learned that there was doubt whether it would get past the Journal peer reviewers and be published. It was published, after minor alterations in 2002 in the *International Journal of Experimental Pathology (London)*. I also did a complete write-up of all experiments which was submitted to Dr. William Eaglestein MD, the Head of the Department of Dermatology and Cutaneous Surgery for his perusal. I did not meet with him for a discussion which graphically illustrated the lack of interest or understanding of my work. I left Miami in November 2001.

In later years, I travelled twice from England to Miami at my own expense to hold meetings with senior University staff to fully declare my deciphering of my experiments, but the University still declined further participation. Under no duress, I was issued an official letter which plainly stated that all Intellectual Property rights belonged to me and me alone. Friends and even UM colleagues found this institutional attitude hard to fathom. I will remain forever grateful that I was offered the chance to pursue my research ideas. America still gives a person (in my case, a non-citizen), a chance to succeed by offering an open mind, generosity of human spirit and 24/7 access to a locked major research building (after hours). I would have had no chance in England with its rigid research structural hierarchy and my having no experience whatsoever with human skin cell research. I would have been turned away.

For the next two years, 2002-2004, I was on a pattern hunt. My instincts told me my work was important. Had the University offered me a job, they would have been the owners of the patents I later filed. I felt I was elbowed aside because someone there had realised the experimental importance of my cellular experiments or had they?

Post-2002, I had also decided to work on the theoretical aspects of my experiments. I did not have access to the cellular photomicrographs locked in a UM file, I could not open. At this time, I became friendly with the owner of a small computer shop in Wigan who was interested in science. He allowed me free access during shop hours to search the internet which I greatly appreciated. I would use this facility for years to scour the published literature before the local Wigan library obtained computers for public use. There, only an hour's use was allowed per day.

I read the reference book on patterns by Philip Ball entitled: '*The self-made Tapestry: Pattern formation in Nature*' published by Oxford University Press in 1999. England being England, one of the national TV channels was show-ing the 007 film, '*You only live twice*'. I found myself perusing the patterns in the opening credits with intense interest while Nancy Sinatra sung one of the loveliest James Bond film themes.

The granting of my UK patent in 2004, the same one I had shown to UM staff members raised my spirits. I was caught in a minefield in no man's land. I was not even a voice in the wilderness. I had no voice. I had no personal funds to attend conferences to present my work, even when invited. Both medical and scientific researchers carried on with their same old theme of using the Harvard technique to grow keratinocytes and published reviews were regurgitations of old, worn themes. Some research academics were promoted to Professors, based on their regular scientific publications which I now considered to be erroneous.

Taking up a research post in Paris, France in 2005, I had decided on a whim to take my UM manuscripts with me in my luggage. I found Paris relaxing and my local bar became the bar opposite the Folies Bergere on rue Richer in the 9th arrondissement. It was just a short ten minute walk from my digs in the 10th arrondissement de Paris. I am not a drinker. I used to drink coffee there regularly. The staff was young and friendly. One asked me to accompany him to an internet cafe for him to communicate with a lady American tourist with whom he had a fling on her visit to Paris. His written English was not good, so I found myself writing (on his behalf) to his American lover telling her that I loved her!

My landlady taught classes in Lyons during the week and on some weekends retired to her country home in Brittany. Therefore, I had the spacious apartment in Central Paris to myself on occasion. On a particular Saturday lunchtime, I was at home and not through choice. Looking through the window of my bedroom, I could see heavy sleet coming down outside. I decided not to brave the elements. Over the weekends, I was doing a lot of walking exploring the Parisian neighbourhoods; it had been twenty eight years since my initial visit. Instead, I dug out my UM manuscripts and sat down at my desk to read them. It was there that I realized the importance of the 'autoengineering' experiment and its fit in the overall scientific puzzle. The experiment was showing on a cellular level the regeneration of the epidermis and was explaining that any bits of complete epidermis left behind, for example, bordering a wound site had to transform itself into (what I privately called) a multilayered island to mimic the stack of epidermal brown rosettes I had seen in the disassembly of the epidermis in warm DPBS after Dispase digestion.

What was a massive discovery was when I realised that only epidermal basal layer rosettes attached to the prepared ECM and in one photograph an infantile basal cell lesion was seen in the shedding of the cell load of an inverted basal layer rosette onto the prepared ECM. This all came about by happy accident. I was idly fingering the smooth surface of a green leaf and I inverted it to see a configuration of veins. The realization that some of the basal rosettes were 'upright' and some 'inverted' dawned on me. The inverted ones were showing a configuration of cells shed from the rosette onto the prepared ECM which reflected the full complement of epidermal cells in the basal layer of the human epidermis. What confirmed my deduction was the Abstract of an online article (in a seemingly obscure Russian journal) by Spichkina OG, Kalmykova NV, Kukhareva LV, Voronkina IV, Blinova MI, Pinaev GP (2006), '*Isolation of human basal keratinocytes by selective adhesion to extracellular matrix proteins*'. Tsitologiia **48**: 841-7. It stands to reason, (my mind declared) that the epidermal brown rosettes adhering to the prepared ECM had to be basal layer ones. I found this Abstract online in one of my periodic searches of the published scientific/medical literature in 2007.

The road widened, because now I realised that I had to seek support in the published literature on integrins (Cell receptors). I believed that the apical cell receptors on the 'upright' basal epidermal brown rosette had to be similar or identical to those on the 'inverted' basal ones for both types of rosette to attach to the prepared ECM in an identical fashion, allowing them to shed their cell loads without hindrance. After a long online search, I found a single paper published by a London group which backed my assertion: Bishop LA, Kee WJ, Zhu AJ, Watt FM (1998), '*Lack of intrinsic polarity in the ligand-binding ability of keratinocyte beta1 integrins*'. Exp. Dermatol., 7: 350-61.

In searching the scientific/medical literature, it is a given that the title of a published paper may not contain that vital nugget of information that is being sought. On the internet, only an Abstract of the paper can be given. There is a fee to be pre-paid before gaining access to read the full article. USA rates can vary between $29-39! Some academic Journals do not allow free reading of articles which date back to 1978 in 2015! I have held the opinion that if a paper has been in the literature for 15 years, it should be freely available for perusal. I was determined to find Journals with an affordable publication fee and free online reading when I decided to write and publish my articles. I thought that would be a personal, humanitarian gesture considering the nature of my work.

Academic journals do not take into consideration independent scientists who want to publish. Some high profile Journals charge authors over UK £1000-. This fee is normally met by the University department or research grant funds. Medical Journal publication fees are always higher still.

In the late 1990s, while in Miami before contacting Dr. Raines, I had burned myself on the heated filament of a domestic oven in my apartment. I did not have a microwave oven and was putting a plate of food on an oven shelf. I was stupidly not wearing my oven gloves. To my consternation, the wire shelf was not seated properly and it tilted! In a desperate effort to save my plate of food, my right hand gravitated upwards and brought the first web space (that is the dorsal area of the hand between the right thumb and the right index finger) in contact with the heated filament. I obtained a painful burn which can be described in medical terms as a second degree burn; one with a raised epidermis and bleeding. I looked around my almost empty bathroom cabinet and found some pure cornstarch and a triple antibiotic ointment, both sold over the counter at my neighbourhood drugstore. After rinsing the burn area with cold water under the tap, I applied both and covered with some ordinary white gauze. To my astonishment, within six weeks it healed properly without any discoloured skin at the site of the burn. This memory would come back into focus more than 12 years later, when I realised that the cornstarch granules were somehow mimicking the epidermal brown rosettes structurally and could be used as biodegradable scaffolds for skin tissue regeneration of the epidermis. Secondly, the epidermis was so thin that a full strength medically prescribed antibiotic was not suitable for topical wound healing efforts. My lightweight triple antibiotic ointment, drugstore-bought was quite sufficient to ward off any minor bacterial infection caused by damaged skin. Broken skin would allow bacteria normally present on the skin surface to gain access interiorly into the wound site. I then learned that *'contact allergies'* can be experienced by some patients when full strength medical antibiotics are topically applied to the skin.

I also realised that cultured skin substitutes had no source of nutrition. These were commercially manufactured cellular skin constructs currently being used in wound healing. The cornstarch granules were carbohydrates, a source of sugar which would provide a ready source of nutrition. To my mind, the reasonable and rational explanation for the application of honey to wounds since ancient times now seemed eminently sensible, especially after it was published that some honeys contained a small, natural antibiotic content.

The Workings Of The Mind

It is well known that there is no scientific explanation for tunes or songs that suddenly impinges in a person's head. They can stay around for a full day, I am told. One day during my tenure with Dr. Berenson in New Orleans, I suddenly found the Beatles song *'If I fell'* in my head. I hummed it for the whole day. It was a period when I left the radio on overnight. The station of choice was WTIX, a well known New Orleans AM station that played the golden oldies as their jingles advertised. I fell asleep. Out of a deep sleep, I found myself awakened as the first stanza of *'If I fell'* sounded on the radio. My mind recognized the song and registered it and I fell back asleep. I have no explanation for that occurrence. It was a one-off. After my tenure in New Orleans, I did not sleep with a radio on. Does a particular piece of music create a resonance on the subconscious mind? Where does that piece of music come from? Deep within the memory bank of my mind? What is the function of the subconscious mind?

A later episode may provide a signpost to an understanding. As I have stated previously, I had a colour photograph on photographic paper of a basal epidermal brown rosette shedding its cell load on a pre-prepared ECM and creating a new micro pattern. I had pinned it to my bedroom wall in Wigan, England after I returned home. In January 2002, dressed and ready to go out to the local shops, I found myself being attracted and tracing the circle of the micro pattern (without actually touching) while involuntarily humming the first two lines of lyrics of the song, '*The windmills of your mind*', an Oscar-winning song penned in English by Alan and Marilyn Bergman (music by Michel Legrand) for inclusion in the sound track of the Hollywood film entitled, '*The Thomas Crown Affair*'. As a young man when I first heard the song, I was intrigued by its lyrics and over the years, it had become a personal favourite. Interestingly, I later learned that the Bergmans thought the lyrics were representative of a 'mind trip'. Where that song came from that morning and into my mind remains a mystery. I had not been humming before my involuntary reproduction of the two lines of lyrics. I cannot reproduce them here otherwise I will be liable for an annual copyright fee after seeking and possibly gaining permission for their use.

Nine years later in 2011, I was deeply concentrating on a reply I was writing to the UK Patent Office when I suddenly remembered that episode. In my mind, macro patterns were created from the two lines of lyrics and when applied to my work under the microscope, the congruent resonance had to imply micro patterns. A micro pattern I concluded had to have some bio-mathematical inference. I immediately looked up Alan Turing's seminal paper on chemical morphogenesis published in 1952 (Turing AM, '*The chemical basis of morphogenesis*'. Phil. Trans. Roy. Soc. B., **237**: 37–72). This now famous, but an unknown back in 1952, brilliant mathematician and code breaker in England during World War II, had proposed a reaction-diffusion mechanism for biological pattern formation providing a biological explanation for animal coats, feather buds and fish skin patterns.

A paper which had registered and lay on my mind as a constant academic niggle for some years was a description of the human epidermis as a mathematical functional unit with a *phi* correlation (Hoath SB, Leahy DG (2003), '*The organization of human epidermis: functional epidermal units and phi proportionality*'. J. Invest. Dermatol. **121**: 1440-1446).

These researchers had no photographs of cellular structures or intra-epidermal cellular structures of the epidermis to support their calculations. I had the intra-epidermal cellular structures which were circular and gave credence to their *phi* (1.618034) conclusion. Their paper now fitted in with my personal thesis that there was a bio-mathematical formula for the structure of the epidermis. Of note is that no medical school includes biomathematics in their curriculum.

Another paper which had stayed on my mind was Zeng W, Thomas GL, Glazier JA (2004), *'Non-Turing stripes and spots: a novel mechanism for biological cell clustering'*. Physica A., **341**: 482 – 494, which in my mind could go some way to explain the cell clustering giving rise to an epidermal brown rosette. Of note, was that Prof. Glazier's paper could not be found on the widely used (USA National Institutes of Health) NIH online search engine, www. PubMed.gov (The National Library of Medicine, USA); only by Googling Dr. Glazier's name, did I find his paper. I personally regarded all three papers listed above as *avant-garde* science which needed careful interpretation by me. It really did take a few months before I was in an informed position to comment on them and included them in my JDBTE 2012 paper. There was a massive chasm between what I personally believed in terms of my interpretation of my cellular photomicrographs and what I could present to the wider scientific/medical community as justification for my thoughts. The three papers described above, published by senior distinguished scientists, taken together, brought me a measure of personal relief as I could cite them as published evidence of a theoretical nature backing my conclusions.

Therefore, I can only conclude that when you start suddenly humming a song out of the blue, your subconscious mind is trying to communicate with your 'live mind' in real time and trying to tell you something. Do not dismiss it as a random occurrence without relevance. Please pay attention.

Continuing On

Dr. Walter Quevedo at Brown University had theorized since the 1960s that the epidermal-melanin unit was part of an organizational unit composed of keratinocytes and the melanocyte in the stratum basale of the epidermis. He had retired and was ailing when I got him on the 'phone and I arranged to send my intra-epidermal micro patterns for him to see. It was shown by other researchers that each melanocyte found in the basal layer of the human epidermis was surrounded by 36 keratinocytes. Each melanocyte looks like, in layman terms, a miniature octopus lying on the basal layer and extending its tentacles to the keratinocytes in the epidermal layers above. Those keratinocytes are properly termed supra-basal keratinocytes (in other words, keratinocytes above the basal layer). It therefore made no sense to me that some senior researchers used enzymatic digestion to isolate melanocytes and to use cell culture to grow them as a contiguous layer. In my humble opinion, this can only be referred to as 'artificial scientific research' or 'artificial cell culture'.

WRITING PAPERS

In autumn of 2011, I decided to write a paper[3], having now gained a clearer understanding of my cellular experiments. My way of doing things is to sit down with a laptop computer balanced on my lap, throw away the clock, meaning disregard time of day and get on with it. As I got in 'the zone' and I composed, edited and cut and pasted paragraphs, I suddenly realised that the dermal-epidermal junction (DEJ) also represented a micro pattern. I hastened to include it and my cellular photomicrograph of it (See Figure 5 in Appendix). By the end of the tenth day, the paper was completed and I was exhausted. I submitted it in November 2011 to the *Journal of Developmental Biology and Tissue Engineering* (JDBTE) (www.academic journals/JDBTE. org). I was informed the publication fee would be US$500-. I had to pay. It was accepted for publication and published in April 2012. It is freely available online.

The copyset editor of this journal introduced an error and two sentences were jumbled. I sent the Journal several e-mail messages pointing out their error in the proof copy they sent to me, but to my chagrin, the sentences were not corrected. I will correct them here: On page 5 of the published paper, in the second paragraph, it incorrectly reads:

The thickness of human epidermis and the papillary dermis layer is 0.3 mm; a further 0.7 mm down lies the reticular dermis (Sorrell et al., 2004), shows a 'keratinocyte mass' formed in the presence of papillary fibroblasts.

My original sentences in my submitted manuscript were: The thickness of human epidermis and the papillary dermis layer is 0.3 mm; a further 0.7 mm down lies the reticular dermis (Sorrell et al., 2004). Their Figure 6A shows a 'keratinocyte mass' formed in the presence of papillary fibroblasts.

For my readers' benefit, I will state the full reference details:

Sorrell JM, Baber MA and Caplan AI (2004). Site-matched papillary and reticular human dermal fibroblasts differ in their release of specific growth factors/cytokines and in their interaction with keratinocytes. J. Cell Physiol., 200: 134-45.

Sitting outside on a friend's porch in Orlando, Florida, a few months after the publication of my JDBTE paper, the son of his neighbour greeted me. He knew I was, as he put it, 'a university person'. He was carrying a file folder which looked rather weighty. He complained that he didn't like Biology and waved the heavy organizer file folder at me, saying he had too much to read. I asked to see the folder. To my surprise it contained a Biology textbook of the Natural World with colour illustrations. As I idly turned pages while still in conversation with him, I got quite a surprise. Internally, the chloroplast of a leaf contained a stack of thylakoids within a granum (plural: grana) which resemble a stack of coins. I had alluded to a stack of coins on page one of my JDBTE publication to try to explain the disassembling of the epidermis into epidermal brown rosettes. I also alluded to the fact that when a basal brown rosette shed its cell load onto a prepared ECM the cellular photograph showed small circles in a 'V' which as I unknowingly described them, resembled tiny, nascent buds in a floral pattern. I realised that when I was a young student studying the chloroplast, modern instrumentation had not been available and things had moved on, meaning there was now a clear understanding of the internal cellular structures within the chloroplast of a leaf. I concluded that there was a degree of commonality in nature's structural blueprint. The stack of thylakoids present within the chloroplast of a leaf arranged in a column collectively named a granum definitely resembled the stack of epidermal brown rosettes disassembling in warm DPBS, which I saw under the microscope.

My second paper[4] was the result of another happy accident. Taking a shortcut through the baking goods aisle in my local Wigan supermarket, I was greeted by a friend. I stopped to converse. During our chat, my eyes strayed to a packet of fine leaf gelatine sheets sitting on a nearby shelf. After my chat was over, I went to the shelf and examined the product description on the back of the packet. My mind strayed to the treatment of burns and skin grafts. Why couldn't fine leaf gelatine sheets be clinically used? The gelatine sheets are absorbent, cheap to manufacture and to buy and after application to a wound or burn, they would shrivel in a normal airflow. I also thought that colour changing markers denoting yeast, bacteria or fungi could be seeded within to allow nursing staff to immediately grasp if a wound had become infected. I dismissed my thoughts as nonsense but later on deep reflection, I decided they could work. I set about to write a paper and looked around for a journal with a cheap publication fee. I came upon a new Journal entitled the *Open Journal of Regenerative Medicine* (OJRM) (OJRM@scirp.org) which could be freely read online. I submitted my paper and it was accepted. The fee was US$400-. I paid.

As I have stated previously, the photographs of my cellular experiments were on files I couldn't open. On a visit to Miami in early 2007, a UM friend transferred the files on the floppy disks to Microsoft Word onto CD disks. In my quest for research funds, I found myself being invited by a major cosmetic company in Paris, France to make a presentation at their headquarters in the last week of March 2007. I arrived early and was treated to an excellent lunch in their elegant dining rooms. In the conference room where I made my presentation, there was overhead, a large screen suspended from the ceiling on which my cellular photomicrographs were displayed. I had previously decided that I could not disclose the full extent of my work, because I was still trying to decipher things and I could not afford to give away everything for free.

I began my presentation and realised that on looking up, I was see-ing my cellular photomicrographs from a new magnified perspective. Normally, one is looking down at a photograph or through a microscope. There I saw certain things for the first time, so while chastising myself inwardly, I had to keep a straight face and continue with my presentation. The two senior company executives, who I later thought were not hands-on bench scientists, realised that they were seeing new things, but I was an independent scientist and they were in a mindset swayed and concreted over by senior world-wide University Professors and the 'informed opin-ion' of their own outside medical/scientific consultants. I could almost see them thinking, how can he be right and everyone else wrong and why had our much vaunted, well-funded onsite company researchers not come up with this work?

Hence, I got the almost tangible impression that there was a huge cred-ibility gap. I also realised that the nugget of information they gleaned was the non-enzymatic method of inducing cells to detach and consequently expand-ing in culture. I was not granted any funds by them. As I left, tearing off my winter jacket because of the unusually warm temperatures at the end of March, I wondered if years later, they would not come to regret their decision. Had they acted upon what they saw that afternoon, age-group cosmetics for both men and women would have been on the market since, and here I am guessing, 2012. If you collect skin tissue segments of different ages, you can compile a cellular library denoting human ageing. I could not present any material on dermal fibroblasts since in 2007, the CD was still considered lost by me. When I found it years later and published my paper on dermal reticu-lar fibroblasts, I was forced to make critical comment on their own company publications. C'est la vie!

During that visit, a Parisian friend christened my epidermal micro pat-terns of the rosette cell load shedding onto the prepared ECM, as *"Les tourne-sols de Denis"*---Denis's sunflowers. On getting back to England, I spent the weekend hastily compiling a patent application having seen for the first time, and realised for the first time, what some of my cellular photomicrographs were really displaying. On Monday morning, I ran through the hill-top churchyard in Wigan in the rain on my way to the main post office branch where I mailed my patent application to the UK Patent Office. Incidentally, it was that Patent Office correspondence, just prior to granting in 2011, to which I was replying, when I remembered my 2002 episode with the lyrics of "The windmills of your mind".

My experience in Paris in March 2007, solidified and validated my past struggles. There at last was a chink of light. I knew that there was still a lot of work ahead of me. It was not yet time for a congratulatory back slap but the 'City of Light' had shed some on me. Je t'aime, t'aime Paris!

In writing my initial paper in 2011, I realized that a CD had been lost or misplaced, as I have related above. I searched high and low for two years until I found it hidden within a cardboard multi-pocketed folder. I was relieved. It contained photographs of dermal fibroblasts. I had continued to read the published scientific and medical literature. I soon realized that with the knowledge I had acquired over the years from reading other scientists' papers that their methodology used to isolate them was incorrect and my scraping of the dermis revealed that dermal reticular fibroblasts at confluence displayed a signature micro pattern *in vitro* (See Figure 4 in Appendix). My confluent cellular configuration had not been seen or published. The cells would pack tightly together at an angled vertical. The *in vitro* packing configuration at confluence provided a way of differentiating between dermal papillary and reticular fibroblasts. The papillary fibroblasts, using the same procedure would show, in contrast, a horizontal packing configuration *in vitro*. To think I had done this more than a decade ago! I was forced to critically comment on published papers and review articles by senior researchers and their techniques and results contained therein. There was no polite way to state bald facts. Again, the manuscript[5] was submitted to OJRM and yes, again I paid the publication fee.

Chronic non-healing wounds are major medical problems that are very expensive to treat especially in diabetic patients. I had walked past the Wigan Hospital library on Wigan Lane innumerable times on my walks downhill to Wigan town centre. There was a large signpost fronting the building; blue background, white lettering. I had promised myself to go in, but somehow never managed it. One day, as luck would have it, I entered the premises and browsed the bookshelves. I came across an edition of the *British Nursing Journal* and started to read an article on wound beds. In the medical community, there was no real proof that 'dry' rather than 'wet' wound beds was the better choice of treatment; that got me thinking and I decided to write another paper questioning accepted conventional wisdom. I suggested 'autologous cell therapy' using dermal reticular fibroblasts and their ECM to 'concrete over' such wounds and an encrypted app. to prevent medical misdiagnosis. Using cells from the same patient mean no ethical or immunological considerations are applicable nor medical institutional permission is needed. A postage stamp-size skin segment will be required, because the cells can be expanded in cell culture. A punch biopsy will provide sufficient skin. I decided to complete a trilogy of papers in OJRM and submitted it there. It was published in 2014[6].

An overview of the problems with wound healing and burns management from scientific and medical viewpoints.

Scientific: (1) Mixing epidermal and dermal cells and trying to co-culture them.

(2) Culturing melanocytes as a contiguous layer, when it was known that each melanocyte is surrounded by 36 keratinocytes.

(3) Blind acceptance of the *Rheinwald and Green technique for cell culturing of keratinocytes as the gold standard for decades. *This was the mistake I identified* (referred to on page 12).

Growth of these *single* cells led to the cell culture of populations of uniform, single keratinocytes with an artificial 'fishnet' morphology with no chance of identifying suprabasal keratinocytes or the epidermal brown rosettes in the basal layer.

(4) Even when a microtome was used to cut thin slices of dermal skin, they were never carefully scraped with a surgical scalpel and blade.

(5) Standard cell culture techniques of pelleting, resuspending and finally plating cells on isolation were incorrect.

(6) No knowledge of the disassembly of a small piece of whole epidermis; too much emphasis on the keratinocyte; there were even specialist keratinocyte laboratories.

(7) Published papers suggesting a bio-mathematical relationship to the epidermis; either ignored or total unawareness of these publications.

(8) No standardized technique for enzymatic digestion of epidermis (including time and temperature to be used).

(9) The dermal fibroblasts have no specific cell marker which created a genuine problem. The use of enzymatic digestion on minced dermis using different times and temperatures (yielding a mixture of both papillary and reticular fibroblasts which cannot be separated), after published reports of two types of dermal fibroblast is inexplicable as was the continued use of the generic term 'dermal fibroblasts' in published papers.

(10) No appreciation of the ECM whatsoever; only the cells mattered.

(11) Cell culture of human cells was relegated to the technician level of laboratory work which did not help academic progress.

(12) Fresh pieces of skin tissue from an operating theatre are highly recommended for the isolation of all skin cells. Frozen segments should not be used. Cells from cadaver skin tissue seem to have lost their 'supple morphology' (this is hard to describe in layman's terms). A simple but recognizable analogy might be comparing the eyes of a freshly caught fish to one stored for several days.

(12) Journal referees who review papers for publication should insist that all methods of cell isolation are clearly stated and cellular photomicrographs of used cells are submitted. A sentence in a published paper which reads, 'normal human fibroblasts et cetera' is meaningless. It should have been red flagged and referred back to the authors for clarification. Questions arise:

What kind of fibroblasts? Papillary or reticular? How were they isolated? Enzymatic or non-enzymatic methodology? What is being inferred by the word, 'normal'? As a result of my work, cell lines seem not to matter. Primary cells straight from isolation, commonly termed P_0 cells are the ones that really matter.

(13) Going forward, biomedical scientists should become more aware of Bio-mathematics and take the subject on board; it should not be regarded as being strictly reserved for mathematical modellers or bio-mathematicians.

Rheinwald JG, Green H (1975). 'Serial cultivation of strains of human epidermal keratinocytes: the formation of keratinizing colonies from single cells'. Cell., **6**:331-43.

Medical: (1) No medical school includes cell culture or biomathematics in their curriculum, which means that even board- certified dermatologists cannot split a skin biopsy into its epidermal and dermal cells after dissecting off the adipose tissue (fatty tissue) in a cell culture laboratory.

(2) Skin grafts even when they 'take' do not allow the regeneration of the dermis and leave behind painful donor sites on the patient's body.

(3) The use of heavyweight bandages pressing down on a wound or burn may actually impede the natural regeneration of the very thin layer of epidermis.

(4) The recognition that the human body may consist of 'postal code areas' with altogether different specific skin identity sites. Non- clinical requests and unnecessary cosmetic surgery of the face using dermal fibroblasts (which ones?: papillary or reticular?) should be reconsidered and future patients warned of the risks.

(5) The practice of waiting 4 weeks for a flimsy keratinocyte auto-graft, properly termed, 'a cultured epithelial autograft' (CEA) for treatment of a burns patient should be reconsidered.

(6) Consideration should be given to injecting 4% (*v/v*) glucose mixed with saline under the burn site (before debridement) to ascertain whether the damaged tissue can be persuaded (over time), to raise off the unaffected area beneath (in essence, trying to create a 'blister' to inhibit contracture) to allow tissue regeneration to commence and to concomitantly provide a much needed measure of hydration. This could be termed 'outside-in' intervention in burns management.

Additional note: The medical/scientific published literature should not be used as an advertising or promotional platform for products created in spin off companies by senior scientists and medical doctors. Papers should not be accepted for publication which contain no description of proprietary processes.

In summary, my good health allowed me to complete my work. I had to dig deep to acquire a personal faith that at the end of the day, I would somehow come through unscathed. It was never a case of winning or losing. It was being true to myself and following my personal instincts. If I have made a small contribution to the well-being of mankind, I will be satisfied that more than a third of my life has not been wasted.

A thorough grounding in the sciences at Lancaster University, Bailrigg, England gave me a good solid foundation. The road to Miami started at Lancaster. The late Professor John Bevington was extremely helpful and supportive. My PhD supervisor, the late Prof. Charles Phelps (PhD, Oxford) had advised me that to make my mark as a scientific researcher, I must publish six novel papers of the highest quality. Even though you have passed on, Charles, I still fondly recall your bonhomie and your good nature. Charles had always fostered the hope that one of his students would graduate and move on to his alma mater, Oxford University. I became his only student to fulfil his wishes.

Allied with this was my prior pre-medical course at Louisiana State University in New Orleans (LSUNO), now the University of New Orleans, where I was introduced to the fundamentals of medical science. I could not complete my medical degree because I was not a Louisiana State resident, a requirement for medical school entry in those days and neither could I afford the medical student fees. With supreme irony, Prof. Gerald Berenson in New Orleans played a key role in introducing me to human cell culture and caused me to change my research focus going forward. During my tenure with him, he had published a book, '*Cardiovascular risk factors in young children*' containing a critique about feeding snack foods to young children. I paid attention and cut out all snacks. Had America listened to those sage, written words, I wonder whether their current, obesity problem would exist.

At UM, Dr. JK Raines was '*the man of the match*'. Without him, I would not have had access to UM's laboratories. He remains a stalwart friend.

I am still desperately sad that Prof. SL Hsia has passed on. Rest in peace, my friend. Posthumously, UM named a laboratory in his honour.

It is my pleasure to acknowledge UM staff with whom I came into contact during my various tenures there.

Personal thanks and a deep sense of appreciation are extended to those scientists and medical doctors, past and present, whose work I have cited in my published papers. It was my honour and privilege to do so.

If the words, 'epidermal brown rosettes' are searched on 'Google', there is a link that leads straight to my JDBTE 2012 paper. I had nothing to do with that.

I continue to find peace of mind in the GESU Roman Catholic Church whenever I visit the city of Miami. Finally, a few words from Tennyson: '*More things are wrought by prayer than this world dreams of*'.

Appendix

REFERENCES FROM MY SCIENTIFIC PUBLICATIONS generated from my 2001 summer in Miami. Reference 1 published in 1992 is of course, excluded; however, it was the starting point of this project. Also references 4 and 6 which were my own ideas, more than a decade later, 2012-2014 as was explained in the manuscript. They are freely available online for anyone to read. For IJEP papers, use www. PubMed.gov and D.E. Solomon as author.

[1]Solomon, D.E. (1992) The seeding of human aortic endothelial cells on the extra-cellular matrix of human umbilical vein endothelial cells. *International Journal of Experimental Pathology*, **73**, 491-501.

[2]Solomon, D.E. (2002) An *in vitro* examination of an extracellular matrix scaffold for use in wound healing. *International Journal of Experimental Pathology*, **83**, 209-216.

[3]Solomon, D.E. (2012) Mimicry of a natural, living intra-epidermal micro pattern used in guided tissue regeneration of the human epidermis. *Journal of Developmental Biology and Tissue Engineering*, **4**, 1-7. (www. academicjournals/JDBTE.org).

[4]Solomon, D.E. (2013) A paradigm for a skin graft substitute. *Open Journal of Regenerative Medicine*, **2**, 28-30. (OJRM@scirp.org).

[5]Solomon, D.E. (2013) Human dermal reticular fibroblasts at confluence display a signature micro pattern *in vitro*. *Open Journal of Regenerative Medicine*, **2**, 99-105.

[6]Solomon, D.E. (2014) Scientific viewpoints with emphasis on dermal cellular regeneration in wound sites. *Open Journal of Regenerative Medicine*, **3**, 22-27.

Cellular photomicrographs taken from my published papers; reproduced with permission. Phase contrast microscopy. All are subject to granted patents.

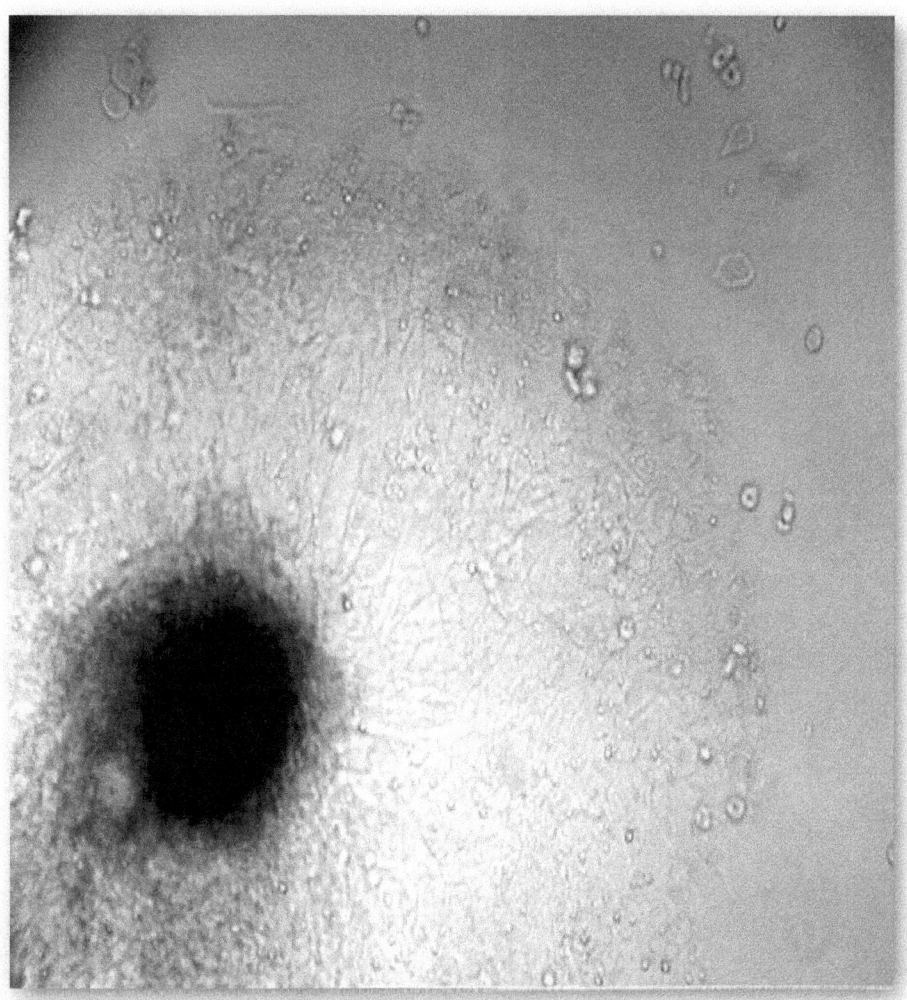

Figure 1. An *upright* attached brown rosette in an upright position displaying its intact micro pattern while shedding its cell load onto a prepared extracellular matrix and creating another micro pattern. Magnification: x 100.
[From reference 3]

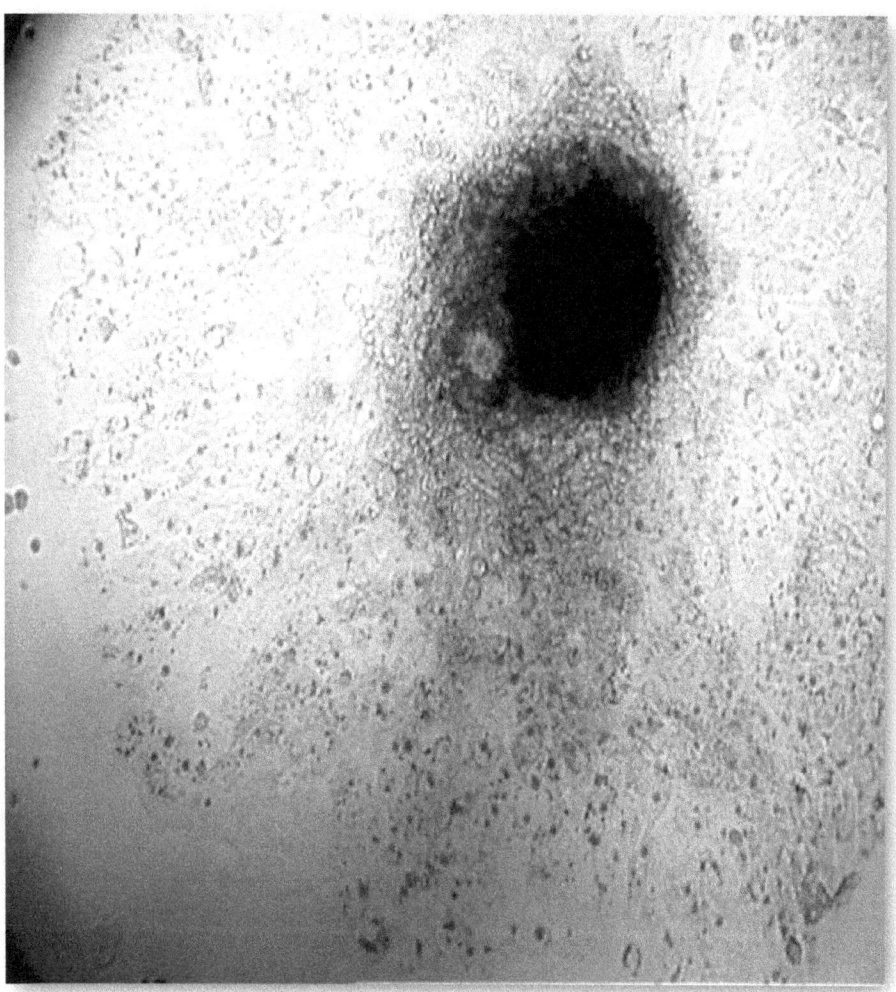

Figure 2. An *inverted* attached brown rosette shedding its
cell load of epidermal cells. Magnification: x 200.
[From reference 3]

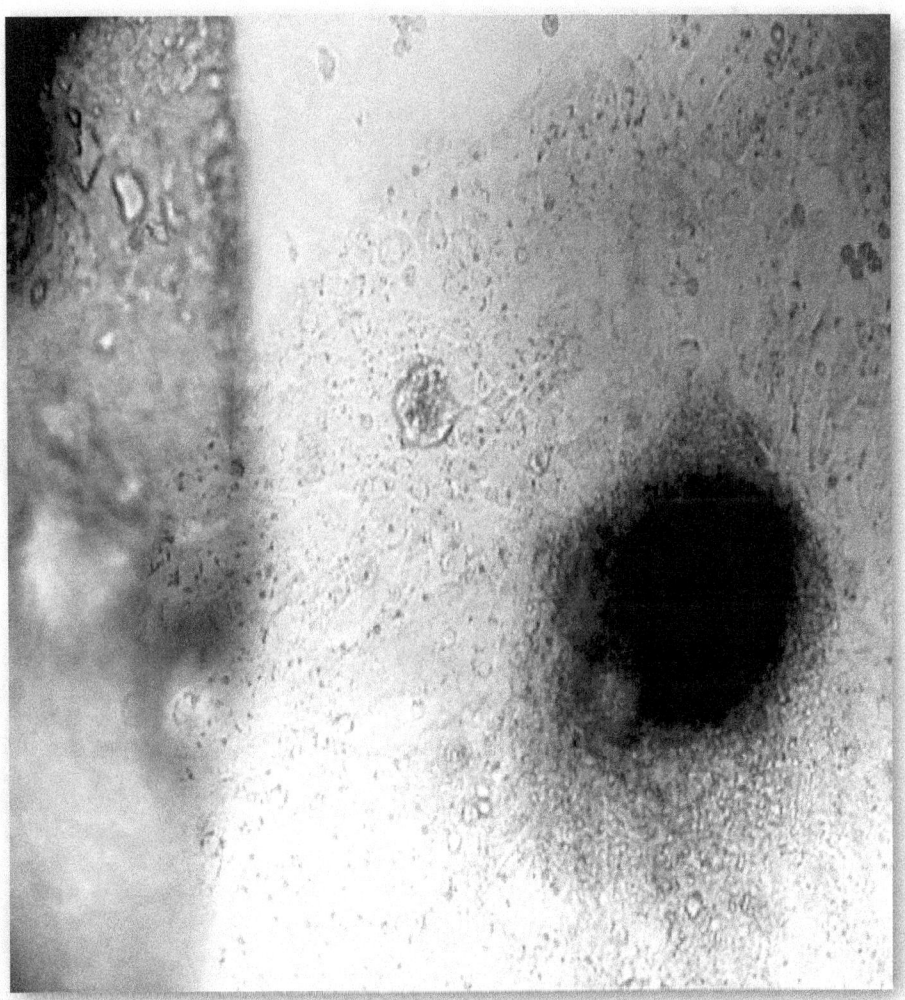

Figure 3. An *inverted* attached basal brown rosette. There appears to be an encapsulated basal cell lesion showing a cellular configuration within. To the left are the edges of the plastic tissue culture dish. Magnification: x 200.
[From reference 3]

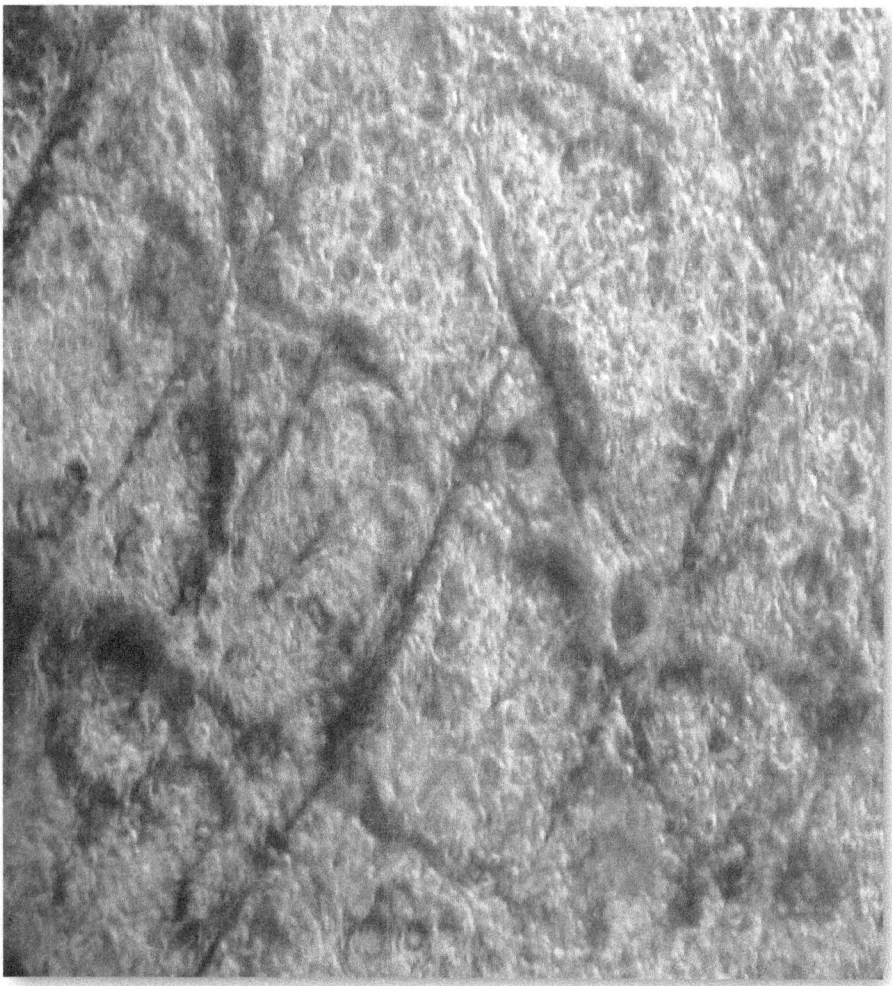

Figure 5. Another micro pattern. The epidermal-dermal junction (after stripping off the Dispase-digested epidermis) displaying dermal papillae and sweat ducts (the 'round holes'). Magnification: x 200.
[From reference 3]

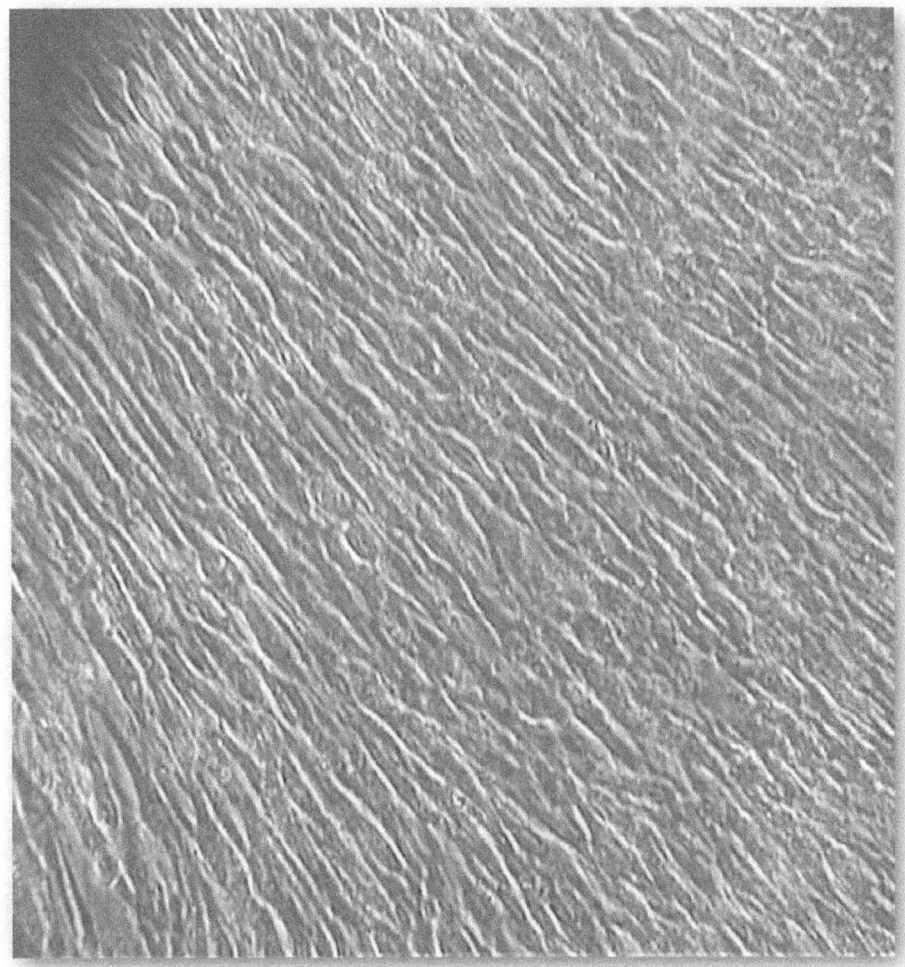

Figure 4. Another view of dermal reticular fibroblasts at confluence. Their side-on signature packing alignment at an angled vertical is clearly visible *in vitro*. Magnification: x 100.
[From reference 5]

Adriaan van de Spiegel-the first person to study human skin.
(Diagrams: Above and on the next page; Copied from Wikipedia)

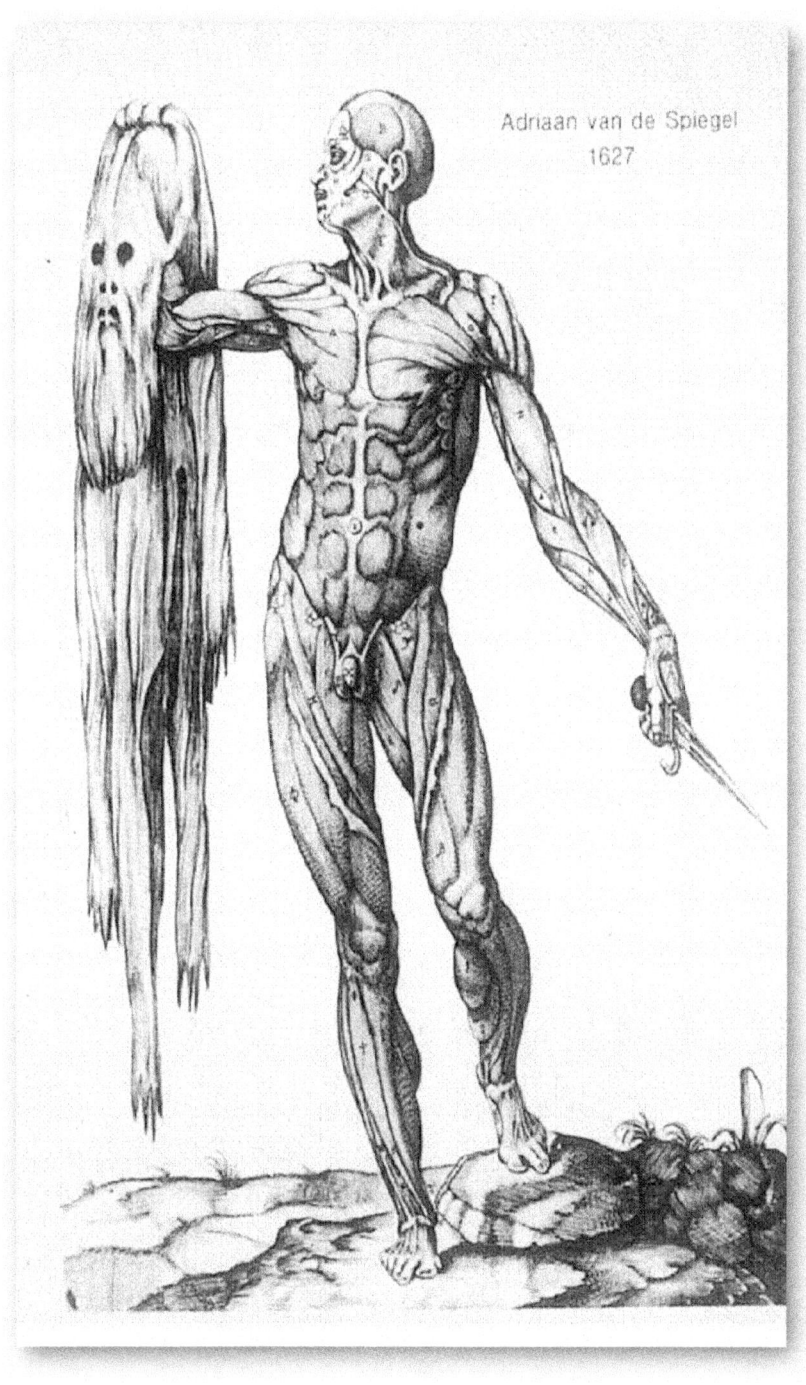

www.ingramcontent.com/pod-product-compliance
Lightning Source LLC
Chambersburg PA
CBHW071001180526
45168CB00003B/1238